Torbay
tī kōuka

Torbay tī kōuka

A New Zealand tree in the English Riviera

WAYNE BARRAR

UPP
University of Plymouth Press

Paperback edition first published in
the United Kingdom in 2011 by
University of Plymouth Press,
Roland Levinsky Building, Drake Circus,
Plymouth, Devon, PL4 8AA, United Kingdom.

ISBN 978-1-84102-297-0

A CIP catalogue record of this book is available from the British Library.

Works from this publication formed the basis of an initial exhibition *Torbay tī kōuka* at McNamara Gallery Photography (Whanganui) October 8–28, 2011.

Designer: Anna Brown
Image manager: Peter Miles

Printed and bound by Printlink, Wellington, New Zealand

UPP
University of Plymouth Press

MADr
WITH
PLYMOUTH
UNIVERSITY

LAND/WATER
& THE VISUAL ARTS

MASSEY UNIVERSITY
COLLEGE OF CREATIVE ARTS
TOI RAUWHĀRANGI

McNAMARA GALLERY
photography

contents

foreword	6
in a New Zealand landscape	8
temperate wetland to tropical effect PHILIP SIMPSON	18
in an English landscape	24
Wayne Barrar's Torbay tī kōuka PETER SIMPSON	58
biographical notes	65
acknowledgements	66

foreword

Artists research people, places and circumstances, often drawing upon local stories and archive materials to inform and contextualise visual explorations. In *Torbay tī kōuka* Wayne Barrar set out to make landscape-based images that indicate the extensive presence in South West England of one of New Zealand's most iconic indigenous plants, *Cordyline australis* – referred to in New Zealand as the 'cabbage tree' or 'tī kōuka', but known in Devon as the 'Torbay Palm'.

Torbay in South Devon runs from Berry Head and the fishing port of Brixham, via the towns of Paignton and Torquay, up to the Ore Stone that marks the northeast point of the bay. Nicknamed the 'English Riviera' because of its mild climate and attractive beaches, it is a long-established British seaside resort. It comes as a surprise, therefore, to learn that 'Torbay palms' are actually from New Zealand, reverse migrants brought to the UK by returning travellers. For visitors from New Zealand, these trees offer a reminder of home – historically this included troops stationed in the UK during WWI.

The Faculty of Arts, Plymouth University, was pleased to host Wayne Barrar's residency in Torquay (March–May 2011). The project and this publication contribute within a series of initiatives linking photography and research at Massey University, New Zealand, and Plymouth University, UK. This was also the first in a new series of international visiting artist-researcher residencies sponsored by the research group for Land/Water and the Visual Arts at Plymouth.

Liz Wells
Professor in Photographic Culture
Plymouth University

Convener
Land/Water and the Visual Arts

October 2011

Cordyline beach edge,
St Ives #1,
England 2011.

in a New Zealand landscape

Point, Western
Lake Wairarapa,
New Zealand 2008

Line of cordylines
and clearance,
Ruamahanga Cutoff,
New Zealand 2011

From lakeside cordyline,
Lake Wairarapa,
New Zealand 2011

Church, Southern
Wairarapa,
New Zealand 2010

Shadow cordyline,
Petone, New Zealand
2010

Regrowth and
forestry on the road
to Wai-O-Tapu,
New Zealand 2010

Roadside cordyline
near Whanganui,
New Zealand 2010

Temperate wetland to tropical effect

PHILIP SIMPSON

Courtesy: *Biodiversity Heritage Library and Missouri Botanical Garden.*

There is a group of trees that have fascinated botanists, gardeners and artists for centuries. Loosely, they are called 'tree lilies', a name coined in German literature at a time when these trees were grouped together in the taxonomic grab-bag, the lily family. In North Africa and Asia they are Dracaenas (the most famous being the Dragon's Blood – the red resin of which Antonio Stradivari used to varnish his violins). In South Africa they are aloes, in America yuccas, in Australia grass trees, and in New Zealand cabbage trees. What they have in common, in contrast to real lilies, is wood. Wood allows them to grow trunks and branches and to become real trees, albeit fleshy, which, in turn, allows them to inhabit dry places. Tree lily landscapes are often awe-inspiring as thousands of oddities, akin to those seen in a Dr Seuss storybook, march across the deserts and cluster around rocks and rivers. With outstretched arms and tousled heads of leaves, they look like people.

New Zealand's cabbage tree *(Cordyline australis)* is one of the finest tree lilies of all: among the largest (with trunks over one metre thick and canopies reaching 20 metres in height). It is most attractive (with large heads of leaves that dance and clatter in the wind, and spring explosions of deliciously scented flowers) and the easiest to grow (in the wet and dry, the subtropical and the subantarctic). For these reasons, cabbage trees have been carried to most countries in the world, and if they can't be grown outside, they grow in pots, indoors.

In New Zealand, their only native country, cabbage trees prefer fertile lowland wetlands, where they grow with other botanical icons, flax *(Phormium)* and toetoe *(Cortaderia)*. Hence, populations can reach landscape proportions in river valleys, along sand-blocked coasts, on wet hillsides, and in forest openings where they form a nursery for the later giants. But their fleshy wood and tough leaves also mean that cabbage trees can grow

Dracaena australis. White-flowered New Zealand Dracaena. *Curtis's Botanical Magazin[e]* v.55 [new ser.:v2] (1828)

almost anywhere, especially in dry hill country like Hawke's Bay and South Canterbury. Furthermore, their many buried buds mean that they can survive a fire or a drought or wind-storm and even an animal attack, and regenerate from beneath the ground, often growing as a cluster of new trunks.

A cabbage tree can grow from a chip off the axe. Their power is so legendary that there is a Māori lament: 'my loss is not that of the cabbage tree that will return'. The beauty of these trees and their easy manner has made them popular with gardeners, and they have been grown at the farm gate, in the suburban garden, along the fence-line, within the park and on the footpath. They are the stuff of motorway landscapes and of conservationists' restoration. However, their very success has also been their downfall. By growing them away from their natural animal and plant associates, people have exposed cabbage trees to a rampant bacterial disease. Over the last two decades, thousands of trees on farms, in gardens and in towns (that is, in cultivated habitats) across the whole of New Zealand, except the far south, have been dying: leaves wilting, browning and falling, bark shedding and stems rotting. It has been nicknamed 'sudden decline' and it starts from a bacterium injected by a leaf hopper. Both bacterium and leaf hopper are native New Zealanders, usually infecting plants only in disturbed places. Never before has the disturbance been so great as today, from humankind. Cabbage trees are marching reluctantly back to their refuge, the wetlands.

When Māori discovered New Zealand and settled here, they brought with them knowledge of a tropical cordyline that had migrated westwards from New Guinea with their ancestors. Māori knew about the valuable leaf fibres, the sugar in the stems and the power of regeneration. They revered their 'ti' as the harbinger of human spirit, and brought it with them in the waka (canoe). What a surprise when they were confronted with the southern giant! They wove the strongest fibres ever known into shoes, capes, bags, cloaks and ropes. They harvested the young tips (cabbages) as a vegetable and the fleshy stems for sugar. They planted them as reliable markers of important places like river crossings and fishing beds. They selected and bred them as a domesticated food crop. They buried their placentas and their dead in the branches. They named every part and every variety and had hundreds of spirits explaining the growth and ecology. Next to tōtara, tī was the greatest gift of the forest god, Tane.

Gratitude and astonishment was the reaction of the early European explorers. Cabbage was a staple on Cook's voyages and the scourge of scurvy was lessened by eating the growing tips of palms and other leafy hearts, collectively known as cabbage palms. Thus came the new name for tī kōuka.

Early descriptions were eloquent. Sydney Parkinson, on Cook's first voyage in 1769, sketched a cabbage tree beside a Māori house, thereby revealing a close identity, and wrote: "The hills are covered with beautiful flowering shrubs, intermingled with a great number of tall and stately palms, which fill the air with a most grateful perfume." (Parkinson, 1773) John Bidwill, in 1841, suggested "If I can get this tree to England, it will make quite a new feature in ornamental plantations". Ernest Dieffenbach, naturalist for the New Zealand Company, described the tī as forming "jungle on the alluvial banks of rivers" and "with the tree-ferns, palms and Dracaenas which abound in New Zealand, that country resembles one situated between the tropics, and especially the beautiful islands of the Pacific." (Dieffenbach, 1844). Later, the government botanist, Thomas Kirk, who trained at a nursery in Coventry and married a nurserywoman, wrote that the cabbage tree "gives a peculiar palmy feature to the landscape, which is unique and attractive..." and added his horticultural bent in saying: "It is a valuable plant for the shrubbery and for ornamental plantations...and gives a tropical aspect to the scene." (Kirk, 1889). This tropical look was widely established in Britain by 1860.

The Forsters described the cabbage tree from Queen Charlotte Sound during Cook's third voyage (1773) and returned seeds to England, seemingly without success. Charles Frazer, appointed as the superintendent of the Sydney Botanic Garden in 1816, and then as colonial botanist and collector, visited northern New Zealand in 1820 and collected seeds for Britain. These were grown in the 'stove' and glasshouse at the Edinburgh Botanic Garden, where they flourished, and in 1827 flowered and set seed (*Curtis's Botanical Magazine*, 1828), but could not be grown outside. A long-term experimental plantation of cabbage trees from throughout New Zealand has clearly established that plants are genetically more tolerant of cold when derived from high altitude or latitude (see Harris et al., (2004) for references). Although the name 'cabbage tree' is firmly established in New Zealand, botanists Thomas Kirk and Thomas Cheeseman, who first became aware of these plants in Britain, both found it inappropriate. Kirk, later followed by Laing and Blackwell in their famous book on New Zealand plants (1906), suggested that 'palm lily' was more appropriate. This followed the British tradition of calling these trees palms, which itself stems from the exploration days of Cook and others and their reference to cabbage palms. The 'palm lilies' of Britain are variously known as 'umbrella palms', 'Manx palms' (Isle of Man), 'Brighton palms', 'Torquay/Torbay palms', and 'club palms', while in Ireland they are known as 'spikes'.

Cabbage trees occur in most temperate countries, usually planted to give exotic effect beside a significant building, a church, a palace or the headquarters of a formal garden. They impart a sense of antiquity to old buildings. In northern California they have escaped from the garden and are regenerating along the edge of Douglas fir forest, sparking concern that a pest has been unleashed.

There are two regions in Britain where cabbage trees stand out as a major landscape feature: Western Scotland, and the 'English Riviera' in Devon and Cornwall. Where an arm of the Gulf Stream raises the winter temperature just enough, cabbage trees grow from Campbeltown on the Kintyre Peninsula, around the Isle of Arran and Cumbrae, along the adjacent Ayrshire coast and inland to Greenock. The bonanza began in 1823 and reached fame when Lady Campbell's tree provided seed, from 1860. In the Logan Gardens and across Arran they seed profusely and add to the food of birds like redwings. The Arran plants were grown from Christchurch seed (southern toughness!) by David Landsborough, in 1879, and his work was continued by Hugh Fullarton of Greenock. Hugh showed that removing the dead leaves from the tree was harmful, and that the trees could survive winter by tying the head of leaves tightly together to a pole (Simpson, 2000). Nevertheless, the winters are damaging. Robert McAllister of Whitefarland wrote: "...there was a snowfall which lodged in the fronds. This was followed by a frosty N. east wind which froze the snow solid with the result that all the fronds fell off leaving an apparently dead tree over two years – it subsequently revived." (Letter to the writer, 1997). Research in New Zealand has shown that higher latitude forms are genetically adapted to retain their dead leaves and that this skirt reduces stem chilling by frost (Harris, et al., 2004).

The Torquay, Paignton and Brixham area in South Devon became known as the 'English Riviera' in the early 1900s, and still is. It was a call for tourists to visit the mild climate, the coastline and the fine gardens. 'Palms' were a major feature of these gardens. They were initially imported in the 1820s, probably from seed from the

The New Zealand Memorial in London's Hyde Park Corner (Paul Dibble and John Hardwick-Smith), which includes a text narrative (from *Dancing Leaves*) related to the arrival of cabbage trees in the United Kingdom
Photograph: Philip Simpson

Edinburgh plant featured above, and did poorly in the cold.

More sturdy cordylines were imported to southern England by John Standish, a nurseryman from Bagshot in Surrey, and a plant provided by him to Kew Gardens created a sensation when it flowered in 1868. The sturdy varieties soon became a major feature and were widely known as the 'Torbay palm'. The subtropical image was promoted by British Rail and Torquay City Council and the cabbage palm became the logo to encourage people to holiday in the area. Thousands of the trees have been planted along the streets of Devon and Cornwall. Many in Torquay were planted to encourage the rehabilitation of New Zealand WWI veterans. Dracaena Avenue in Falmouth reminds us of an earlier taxonomy.

As in Scotland, periodic freezing kills the elevated trunks and the trees survive from buried stems that sprout new shoots. Hence, there is a constant search for improved hardy varieties adapted to cold, and to create forms and colours suitable for indoor- or boutique-garden culture too. In fact, the purple-brown form is hardier than the natural green, a widespread phenomenon among New Zealand plants.

The cabbage tree has natural variation throughout its range. Sometimes mutations occur and leaves with variegated, brown or red parts are formed. The cabbage tree can hybridise with its relatives in New Zealand and this process unleashes all manner of normally hidden traits. This flexibility has led to an explosion worldwide of cordyline breeding. Furthermore, tissue culture (micropropagation), whereby thousands of plants can be generated from a single mutant, has changed a gentle art into mainstream 'plant barn' business. These variations created for horticultural use are termed 'cultivars'.

There are dozens of cordyline cultivars in New Zealand and other countries, including Britain and France. In fact cordyline breeding began in Europe, in 1870, with the release of 'Lentiginosa', described as having leaves tinted and spotted brownish-red. A very famous cultivar, 'Albertii', with strongly yellow variegations, was a mutant from Belgium found in 1896. A hybrid between a purple cabbage tree and the forest *Cordyline banksii*, resulted in the magnificent 'Purple Tower' and subsequent crossing produced the Pacific series of multi-coloured red forms suitable for container growing. The Torbay Palm Farm in Torquay has been a centre for cabbage tree breeding and several named hybrids have been created there, including Torbay Green, Torbay Red, Red Star, Sundance, Sunset, Dazzler, Coffee Cream and Razzle Dazzle. Torbay Dazzler, a variegated form, received Britain's prestigious Royal Horticultural Society Award of Garden Merit in 2004.

Beautiful they may be, but one has to wonder whether the integrity of the original, inspiring tropical look is being maintained by this tinkering.

Man loading young cordyline plants from nursery trolley, Devon, England 2011 (sequence)

The robustness of cabbage trees that started all this off has gone from our gardens, with tissue culture plants getting 'footrot' and demanding high energy food to maintain suitable growth. Our creations need pampering. And there is a political message. The founding document of New Zealand, the Treaty of Waitangi, guarantees the ownership of natural treasures to the Māori tribes. The Flora and Fauna Hearing, completed in 2011, upholds this view. The horse has bolted, but nevertheless the message is sent, that nature is not just a playground but is to be respected, sustained, managed in as green a way as possible. New Zealanders might feel proud to see the homely cabbage tree on the tourist T-shirt, but the real work is, indeed, at home. Here, tree lilies are reeling from the cultural onslaught, their wetland habitat confined behind the fences and the stopbanks, and it is taking a mighty effort to reverse the flow.

Acknowledgement:
Richard Smart, for information on the introduction of cordylines to Britain.

References:

— Bidwill, J. C.1841. *Rambles in New Zealand*. London, S. Orr and Co.
— Dieffenbach, E. 1844. *Travels in New Zealand*. London, Longman Brown.
— Harris, W.; Beever, R.E.; Parkes, S.L.; McSeveny, T. 2004. 'Genotypic variation of dead leaf retention by *Cordyline australis* (Lomandraceae) populations and influence on trunk surface temperatures', *New Zealand Journal of Botany* 42: 833–845.
— Kirk, T. 1889. *The Forest Flora of New Zealand*. Wellington, Government Printer.
— Laing, R.M. and E.W. Blackwell. 1906. *Plants of New Zealand* (4th ed., 1940). Auckland, Whitcombe and Tombs.
— Simpson, P. 2000. *Dancing Leaves: The story of New Zealand's cabbage tree, tī kōuka*. Christchurch, Canterbury University Press.

in
an
English
landscape

Pink scene with
planting, Shaldon,
England 2011

Union Jack and sole
cordyline, Devon,
England 2011

Cordyline (Torbay
dazzlers) in Paignton,
England 2011

Gorilla fringed in cordyline, crazy golf course, Torquay, England 2011

Tiger in tropical
themed crazy golf
course, Torquay,
England 2011

Interpretive sign on
Rock Walk, Torquay,
England 2011

Garden signage, South
West England 2011

The illuminations #1,
Torre Abbey Gardens,
Torquay, England 2011

The illuminations #2,
Torre Abbey Gardens,
Torquay, England 2011

Opportunistic
cordyline, Cornwall,
England 2011

Cordyline fringed
figure, Plymouth,
England 2011

British bulldog/
cordyline trees,
Torquay, England 2011

Clotted cream
fudge, with Cornish
landscape, England
2011

Topiary and cordylines,
Torquay, England 2011

Front gardens/
residential street,
Torquay, England 2011

Torbay palms for sale,
Devon, England 2011

The illuminations #3,
Torre Abbey Gardens,
Torquay, England 2011

Fourteen properties along Dracaena Avenue (cordyline plantings), Falmouth, England 2011 (sequence)

Patriotic Planting,
signage in nursery,
Devon, England 2011

Through cordyline,
St Ives, England 2011

Residential house
and garden, Torquay,
England 2011

Across phormium
to cordyline/hotel,
Torquay, England 2011

Traffic Island #1,
Penzance,
England 2011

Traffic Island,
Cornwall #1,
England 2011

Tall cordyline and
hydroslide facility,
Paignton, England 2011

Seedling cordylines
in nursery, St Austell,
England 2011

Inflatable seaside products on large cordyline, Dawlish Warren resort, England 2011

Pruned cordyline,
Exmouth, England 2011

Specimen cordylines,
Paignton seaside #1,
England 2011

Specimen cordylines,
Paignton seaside #2,
England 2011

Wayne Barrar's Torbay tī kōuka

PETER SIMPSON

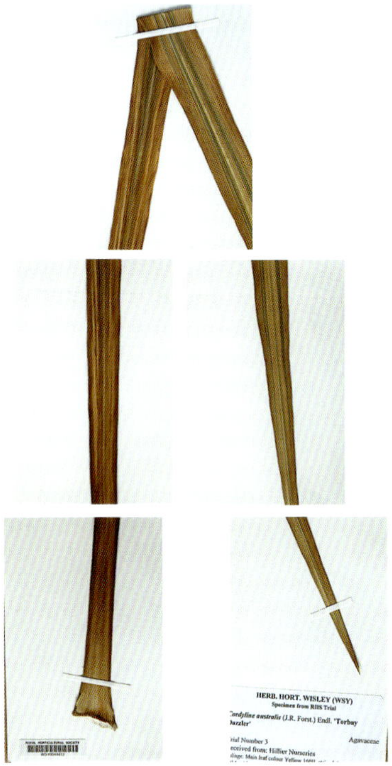

Five Torbay dazzler specimen sheets from RHS Trial, Royal Horticultural Society herbarium, Wisley, England 2011

Recently on a sunny spring Saturday, I drove south from Auckland to the river city of Whanganui on the west coast of the North Island, a distance of some 280 miles, to attend the opening of Wayne Barrar's exhibition *Torbay tī kōuka* at the McNamara Gallery. 'Tī kōuka' is the Māori name for *Cordyline australis*, a tree lily indigenous to New Zealand, popularly (if unromantically) known, ever since James Cook's 1769 voyage, as the 'cabbage tree' because of its edible shoots. I was freshly alert to the presence of cabbage trees in the landscape having spent a week or two looking at a disk of Barrar's portfolio of 'Torbay palms' photographed in coastal Cornwall and Devon, where the species has been established for more than a century.

I wasn't out of the city before the first specimen showed up, a single spindly stem cramped beside a car wash – the sort of marginal urban location Barrar is especially alert to, whether he is photographing in England or New Zealand. Out on the Southern Motorway, the occasional cordyline flashed past in the vegetation lining the highway; hard to tell, usually, whether they were deliberately planted or survivors from a time before motorways were thought of. In the lushly green Waikato dairy country, cordylines seemed less in evidence than clouds of pink cherry and peach blossom near farmhouses or the ubiquitous willows, poplars and oaks showing their tender green leafage. This is a landscape almost wholly made over to exotics, except for moody patches of subfusc-toned kahikatea (white pine) left in a paddock with the occasional spiky cordyline visible around their fringes.

It was quite a relief, south of the King Country towns of Te Kuiti and Taumarunui, to rise and rise up onto 'the roof of the island' (as Frank Sargeson called the Volcanic Plateau) where native vegetation finally takes over and the cordylines visible from the road are the splendid wide-leaved blue-grey *Cordyline indivisa* – the mountain cabbage tree or tōī, not the more familiar *Cordyline australis*. Beyond Raetihi, where the road drops down steeply following the deep gut of the muddy Whanganui River towards the coast, cordylines seemed to vanish from the landscape almost completely, apart from occasional specimens silhouetted on a steep hillside or stranded in a paddock. Nearing Whanganui, though, a magnificent cluster of cabbage trees suddenly appeared by the roadside, close to a farmhouse; there were maybe twenty or thirty mature trees, some in flower, some sited on a grassy knoll, all seeming healthy, not a vestige of 'sudden decline syndrome' in sight. It was a slightly surreal experience, an hour or so later, to walk into Paul McNamara's photographic gallery and encounter the striking images of tī kōuka documented by Barrar as a ubiquitous presence on the south west coast of England.

Many New Zealanders, both Māori and Pākehā, have a special relationship with cabbage trees, as my brother and fellow-essayist Philip Simpson discusses in his definitive book, *Dancing Leaves: The story of New Zealand's cabbage tree, tī kōuka*. My brother and I, for instance, grew up on a dairy farm in Takaka, a narrow steep-sided valley near the north-western tip of the South Island. Each school-day morning we walked for half a mile or so along a gravel road, no houses in sight, to catch the bus to school in the township ten miles down the valley, reversing the journey at the end of the day. Almost all the trees and shrubs we passed on our walk were exotics – the oak, ash, pine and walnut trees near our house, the strip of hawthorn, gorse and barberry bushes between the road and the river – except for a solitary, tall, many headed roadside cabbage tree (as we called it) that was a special marker on our daily trudge, a mark on the map of our young lives. It is gone now, one of the many victims of sudden decline that devastated cordylines throughout the country in the 1980s and 90s (see *Dancing Leaves*, p.91). Probably more than any other single tree, it signifies for me, so far as a tree can, my sense of identity.

It is perhaps because of these special connotations that I find Barrar's photographs of cabbage trees growing on the other side of the world so strikingly disorientating. New Zealanders are used to living in a drastically made-over landscape where most of the vegetation – except in remote or mountainous parts – is imported from elsewhere, especially Britain. But we don't expect the reverse to occur. I felt a similar surprise when I encountered pōhutukawa (a scarlet flowering northern coastal tree, equally iconic) growing along the beachfront in Santa Monica, California. This sense of disjunction, this 'making strange' (as the Russian formalist term *ostranenie* translates) is what gives Barrar's portfolio its special charge, at least to a New Zealander. I can only speculate how these photographs must strike the

Cordyline (in site of initial nineteenth-century planting), Trewidden formal garden, Cornwall, England 2011

eyes and sensibilities of those who live in the UK or elsewhere.

No doubt largely because of its exceptional form – especially by contrast with most of the native plants of New Zealand – *Cordyline australis* has always attracted attention from artists and photographers, beginning with the earliest known depiction of the species by Sydney Parkinson, the artist on James Cook's first circumnavigation of New Zealand in 1769. Parkinson (who did not survive the voyage) sketched a tall specimen with multiple leaf clusters closely adjacent to a Māori whare (dwelling) on Motuarohia Island in the Bay of Islands in the far north.

Subsequently, the inclusion of the distinctive profile of *Cordyline australis* became a regular desideratum in landscape art, largely (though not exclusively, it served other purposes also) as a sign of the exotic in the work of many nineteenth-century artists, whether visitors or settlers, including William Fox, Charles Heaphy, George French Angas, Eugene von Guerard,

John Gully, John Kinder and others.

Kinder, an Auckland-based Anglican clergyman, was a photographer as well as a watercolour painter often choosing the same subject for both media. For example, both a watercolour sketch and a photograph, made in the 1860s, exist of a large named tī kōuka, Te Ti Tutahi, sacred to Ngāti Waiorua, a well known landmark in colonial Auckland. One frequently photographed location, by D.L. Mundy and others, was the so-called Cabbage Tree Swamp at the base of Mount Eden (Maungawhau) – one of the extinct volcanic cones on Auckland's isthmus – a dense forest of cordylines in a wetland later drained and turned into Eden Park, a famous rugby stadium. Other nineteenth-century photographers who depicted cabbage trees were the tirelessly itinerant Alfred Henry Burton, who recorded giant cabbage trees left after forest clearing in the central North Island, and Dr. A.C. Barker, a physician who documented the dwellings, gardens and personages of his fellow Canterbury settlers, who favoured cordylines as a garden tree and sign of the indigenous.

Early last century, the pictorialist photographer, George Chance, made what is perhaps the most frequently reproduced photograph of cabbage trees, *The Storm, Lake Wanaka* (1930), which shows a group of young trees on the shore of the Central Otago lake, their leaf-crowns blown almost horizontal by the violence of the wind. Of twentieth-century painters, those who made most of the cabbage trees as subject matter were Russell Clark, who emphasised by simplification and stylisation the strong patterns formed by the trunks and leaf-crowns, and Eric Lee-Johnson, who initially painted cabbage trees as survivors of forest destruction, and later abstracted elements of their form to signify the life-force, "The force that through the green fuse drives the flower." (as Dylan Thomas put it). Among later photographers, Robin Morrison, Jocelyn Carlin, Patrick Reynolds and Wayne Barrar are among the many who have found something fresh to see and say about this ubiquitous subject.

Barrar's New Zealand photographs of cordylines belong with his broader photographic project, his unique and extensive documentation of the points where nature and culture intersect, where human intervention has modified existing environments through engineering projects, clearance and fencing of land for farming, the planting of exotics for forestry or flood control, the importation of birds and animals, or the myriad other processes by which the land is made-over by human agency. Cabbage trees naturally cluster in such liminal places at the edge of forests, or where land meets water, town merges with countryside, or civilisation with wilderness. *Roadside cordyline near Whanganui, New Zealand* is a case in point; this vigorous specimen, loaded with the aftermath of recent efflorescence, is the only tree in sight as cars speed

Specimen cordylines and park, Teignmouth, England 2011

Container vessel/ coastal cordyline, Falmouth, England 2011

Bungalow on the walk to Daddyhole Plain, Torquay 2011

past on the highway. Similarly, the black and white images taken around the edges of Lake Wairarapa and the associated Ruamahanga Cutoff (so-called because a river was diverted to bypass a lake for flood control) depict the kind of places where cabbage trees flourish and where Barrar likes to position himself and his camera. In *Line of cordylines and clearance, Ruamahanga Cutoff*, the elongated format emphasises the stately row of trees, viewed across the water in which they are reflected, which share the river bank with huge mounds of cleared trunks. As with many of Barrar's photographs, the viewer needs to look beyond the aesthetically pleasing composition to the wider connotations of the image.

Beyond the common and constant presence of tī kōuka which links them together like a string of variegated beads, the photographs taken in Cornwall and Devon, which constitute the bulk of this publication, are too heterogeneous in character to enable simple generalisation. Perhaps the root perception lies in the concept of reverse colonisation or (to borrow a term from the *Star Wars* movies) "the empire strikes back".

Barrar is acutely aware of how the processes of colonialism involve ecological transformation along with social, anthropological and cultural change. As he stated in a 2009 public lecture ('Uncertain Borders: Re-positioning photography of landscape change'), "Both Māori and Pākehā settlement of New Zealand was accompanied by the introduction of plants and animals – some unwitting, some intentional…Where Britain went, so did its ecologies".

In more recent projects, Barrar has gone beyond the initial phase of colonialism – where British influences predominated in the acclimatisation of birds, animals, fish, crops, trees, garden plants and weeds (a process massively documented in that classic of ecological literature *Tutira: The Story of a New Zealand Sheep Station*, by H. Guthrie-Smith, first published in 1921 and revised several times) to consider some ecological manifestations of globalisation. This is also apparent in recent projects documenting the spread of koi carp in certain North Island lakes and streams and the invasion of New Zealand rivers by the pernicious algae didymo, whose effects are well summed up in the popular term 'rock snot'.

Compared to such alarming phenomena, the 'invasion' (if it can be called that) of the warmer parts of Britain by *Cordyline australis* is presented as comparatively benign. At least the citizens of Devon and Cornwall appear to think so, adopting the so-called 'Torbay palm' to enhance the appeal of the region to locals and visitors alike as a sort of sub-tropical playground, as in such seductive images as *Cordyline beach edge, St Ives #1* or *Inflatable seaside products on large cordyline, Dawlish Warren resort* (a title as deadpan as the image itself is scintillatingly colourful). Some images, especially those of dead, dying or severely pruned trees, such as *Bridge edge cordyline, Totnes* or *Pruned cordyline, Exmouth* convey a sense of pathos, of being unloved strangers in a strange land, though just as many others communicate a sense of confident and untroubled assimilation, as in the triptych *Cordyline (in site of initial nineteenth-century planting), Trewidden formal garden, Cornwall*, or the noble sequence, *Fourteen properties along Dracaena Avenue (cordyline plantings), Falmouth*. The preference for using cordylines as plantings on street roundabouts or islands, as seen in *Traffic Island #1, Penzance* and *Traffic Island, Cornwall #1*, plays – consciously or not – on the notion of the 'tropical island', an exoticism more overtly and humorously exploited in *Gorilla fringed in cordyline, crazy golf course, Torquay* and *Tiger in tropical themed crazy golf course, Torquay*.

Earlier generations of New Zealanders sometimes flattered their homeland with the moniker 'Better Britain of the South', an epithet brought somewhat illogically to mind by the association of cordylines with such icons of British identity as the Union Jack (*Union Jack and sole cordyline, Devon*) and John Bull (*British bulldog/cordyline trees, Torquay*). I enjoyed the touches of sly, subversive humour in such images as *Opportunistic cordyline, Cornwall* and the element of poetic suggestiveness in *Illuminations #1, Torre Abbey Gardens, Torquay, England* with its antipodal suggestions of the reversal of night and day, up and down, England and New Zealand. These few observations can only scratch the surface of this endlessly intriguing and thought-provoking portfolio.

References:

— Simpson, P. 2000. *Dancing Leaves: The story of New Zealand's cabbage tree, tī kōuka*, Christchurch, Canterbury University Press.

— Barrar, W. 2009. 'Uncertain Borders: Re-positioning photography of landscape change': public lecture at Massey University. Unpublished manuscript notes.

— Guthrie-Smith, H. 1921. *Tutira: The Story of a New Zealand Sheep Station*. Edinburgh, Blackwood.

From cordyline to
rooftop garden,
Babbacombe Model
Village, England 2011

biographical notes

WAYNE BARRAR

Wayne Barrar is an Associate Professor at the School of Fine Arts, Massey University, Wellington. His most recent solo exhibitions include *An Expanding Subterra* (Dunedin Public Art Gallery, 2010, with venues in 2011 including the City Gallery, Wellington and American University Museum, Washington DC), and *Contact Topographies*, Milford Galleries Dunedin, 2010.

Recent group exhibitions include *Picturing Eden*, a touring exhibition (and major catalogue), International Museum of Photography and Film/George Eastman House, USA, 2006; *Manapouri: Art, Power, Protest*, National Library Gallery, Wellington, 2007; *Land Wars*, Te Tuhi Centre for the Arts, Auckland, 2008; *Toi Te Papa, Art of the Nation*, Te Papa, Museum of New Zealand, Wellington, 2008; *AC/DC: The Art of Power*, Gus Fisher Gallery, Auckland, 2009; and *Community Garden*, Michael Hirschfeld Gallery, City Gallery, Wellington, 2010.

Wayne Barrar's books include *Shifting Nature* (Otago University Press, 2001) and *An Expanding Subterra* (Dunedin Public Art Gallery, 2010).

PHILIP SIMPSON

Philip Simpson is a consulting botanist living at Pohara, Takaka, New Zealand. He was raised on a farm in the Takaka Valley and developed an early love of native flora. He studied plant anatomy at Canterbury University, Christchurch, and obtained a PhD from the University of California, Santa Barbara. Returning to New Zealand in 1976 he worked in soil conservation, environmental education and conservation botany. In 1996, he was awarded the Stout Research Fellowship at Victoria University of Wellington and wrote *Dancing Leaves*, a comprehensive account of the natural and cultural history of *Cordyline australis*. Subsequently, for the Project Crimson Trust, his *Pōhutukawa and rātā, New Zealand's Ironhearted trees*, won the Montana medal for non-fiction in 2005.

Philip Simpson was awarded the Michael King Writers' Fellowship in 2009. He is completing a book on the famous New Zealand timber tree, *Podocarpus totara*. Married to Dr Wendy Parr, a wine scientist, together they manage a boutique winery. Philip has five adult children, six grandchildren, and a flock of feral merino.

PETER SIMPSON

Peter Simpson is Director of the Holloway Press at the University of Auckland, where previously he was Head of English and Associate Professor. He has written or edited about fifteen books, all relating to literature, the visual arts and culture, generally in New Zealand, including two on Colin McCahon: *Answering Hark* (Craig Potton, 2001) and *The Titirangi Years 1953–59* (AUP, 2004).

Peter Simpson's most recent books are on the artist and printer Leo Bensemann, the culmination of a thirty year project, 1981–2011, namely *Fantastica: The World of Leo Bensemann* (AUP, 2011) and (as editor) *Dark Arts: Notes on Printing, Publishing and Painting by Leo Bensemann* (Holloway Press, 2011). Peter has written several photography catalogue and journal essays and contributed a long essay to *Peter Peryer Photographer* (AUP 2009).

acknowledgements

Dawlish town signage among cordyline planting, England 2011

A number of people have provided valuable assistance in the development of this project, for which I am very grateful. Particular thanks must go to the Land/Water and the Visual Arts research group at the University of Plymouth, whose researcher residency made it possible for me to undertake the work in South West England. Liz Wells took a leading role in this regard, and I greatly appreciate the support and feedback I received from her and other faculty members including Liz Nicol, Simon Standing, Jem Southam and Carole Baker.

Thanks, also, to the School of Fine Arts, Massey University, for ongoing support; Jeremy Diggle, for his initial work in developing the Plymouth collaboration; and Heather Galbraith, for advice and assistance in bringing the publication into being.

Torbay tī kōuka is greatly enhanced by the inclusion of the expert essays by Peter Simpson and Philip Simpson; the terrific design work of Anna Brown; and the expertise of Peter Miles in managing the image files. My thanks to all of them, and to Paul McNamara, McNamara Gallery Photography, Whanganui, for his advice and initial exhibition of work from the project.

A number of people in the UK helped with background research. In particular, I would like to thank James Treseder, who provided useful background on the early use of cordylines in the South West, and James Armitage of the Royal Horticultural Society Herbarium at Wisley, who informed me about the development and cultivation of the various varieties and their history.

Thanks again to my partner Kerry Hines for all her help (and GPS navigating along tiny English lanes!).

Wayne Barrar